FOR 'THE TAPPERS'—MY CHEER SQUAD
SINCE THE LARVAE STAGE
-RN-

TO MY DAUGHTER, TARA, FOR BEING MY
VERY KIND CONSTANT, OBSERVANT CONSULTANT
AND DISCUSSANT ABOUT MY ARTWORK.
XX SF

BrilliANT

Text © Rosi Ngwenya 2024
Illustrations © Sandy Flett, 2024

ISBN: 978-0-6458693-0-9

Published by Riveted Press, 2024
rivetedpress.com.au
PO Box 5201
Alexandra Hills QLD 4161
Australia

A catalogue record for this book is available from the National Library of Australia.

Printed in China by Shanghai KS Printing

9 8 7 6 5 4 3 2 1

ROSI NGWENYA
SANDY FLETT

BRILLIANT

EXPECTANT

INFANT

ANT

MILITANT

IMPORTANT

CONTESTANT

TRIUMPHANT

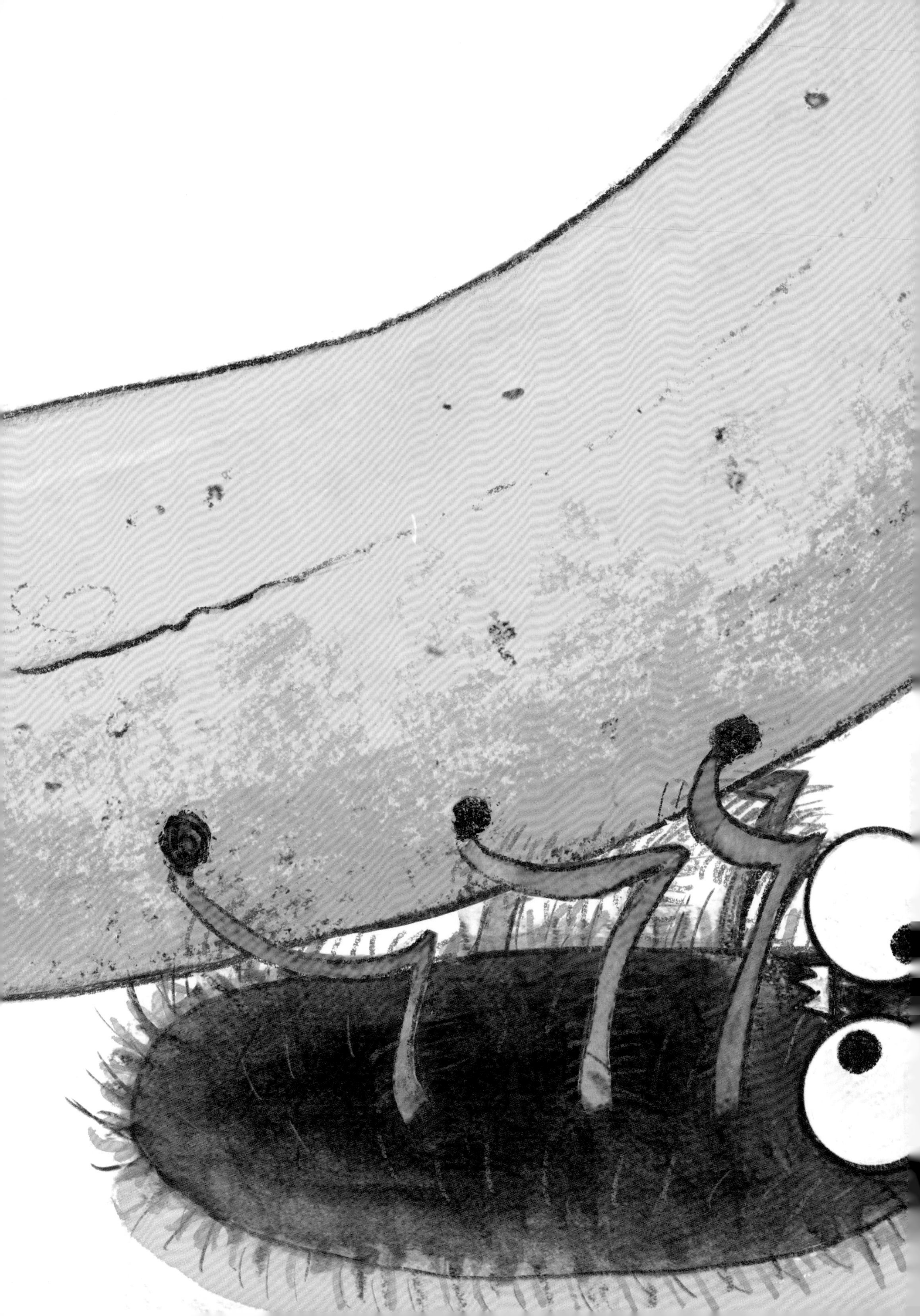

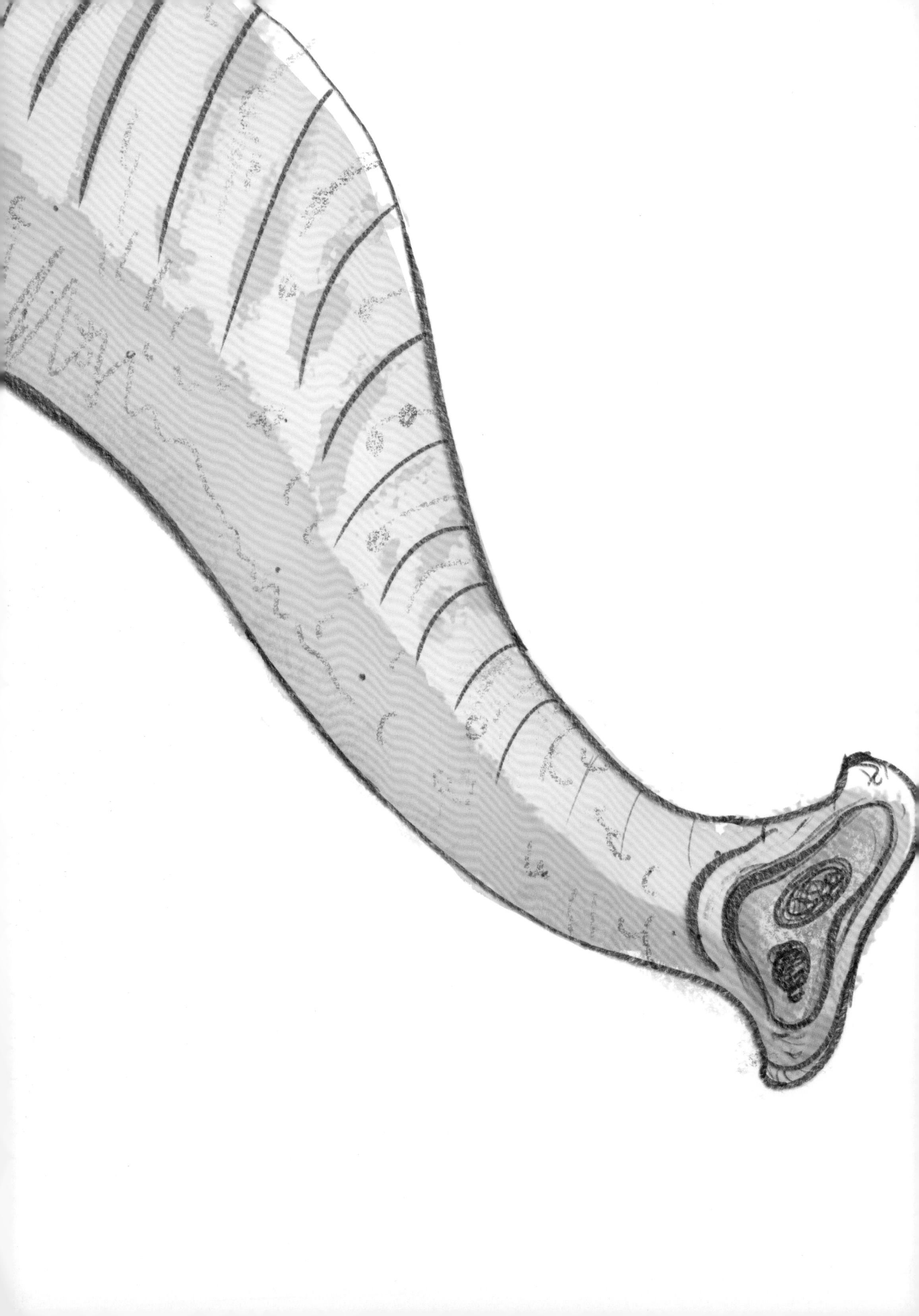

ELEPHANT

WAIT!
WHAT?

ELEPHANT

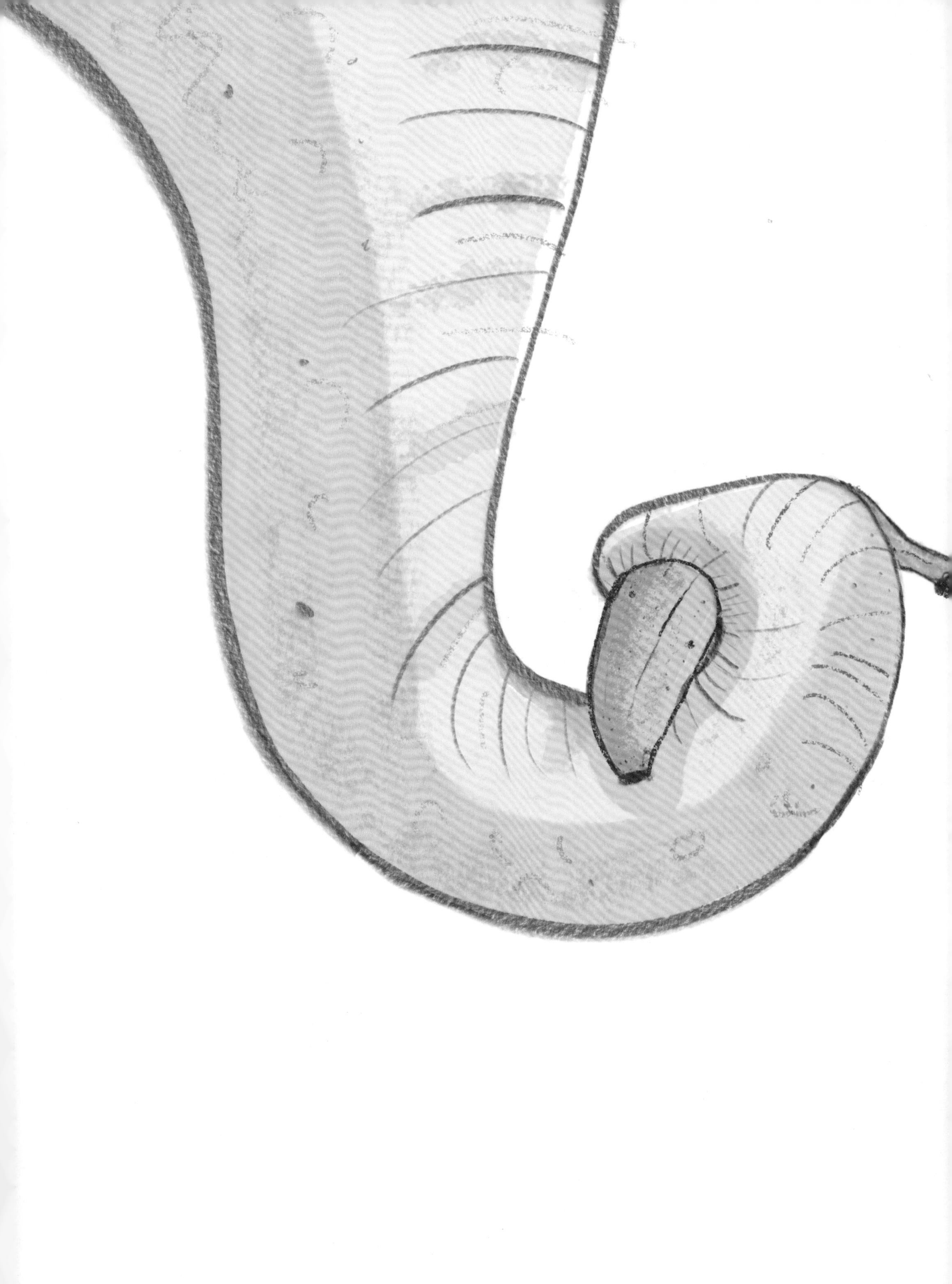

FACEPLANT

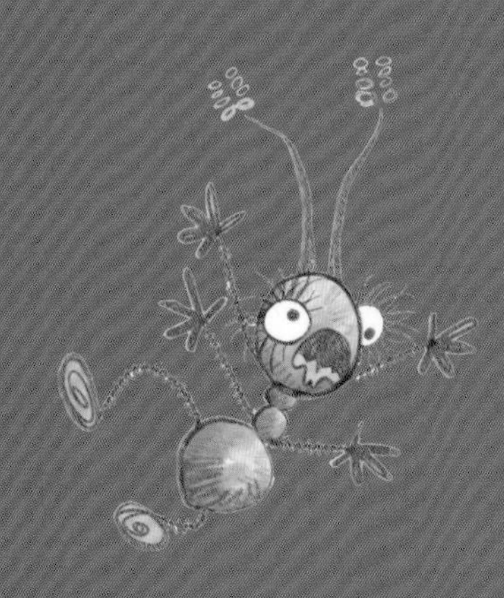

ANTBULANCE

ASSISTANTS

ABUNDANT

OUT for LUNCH

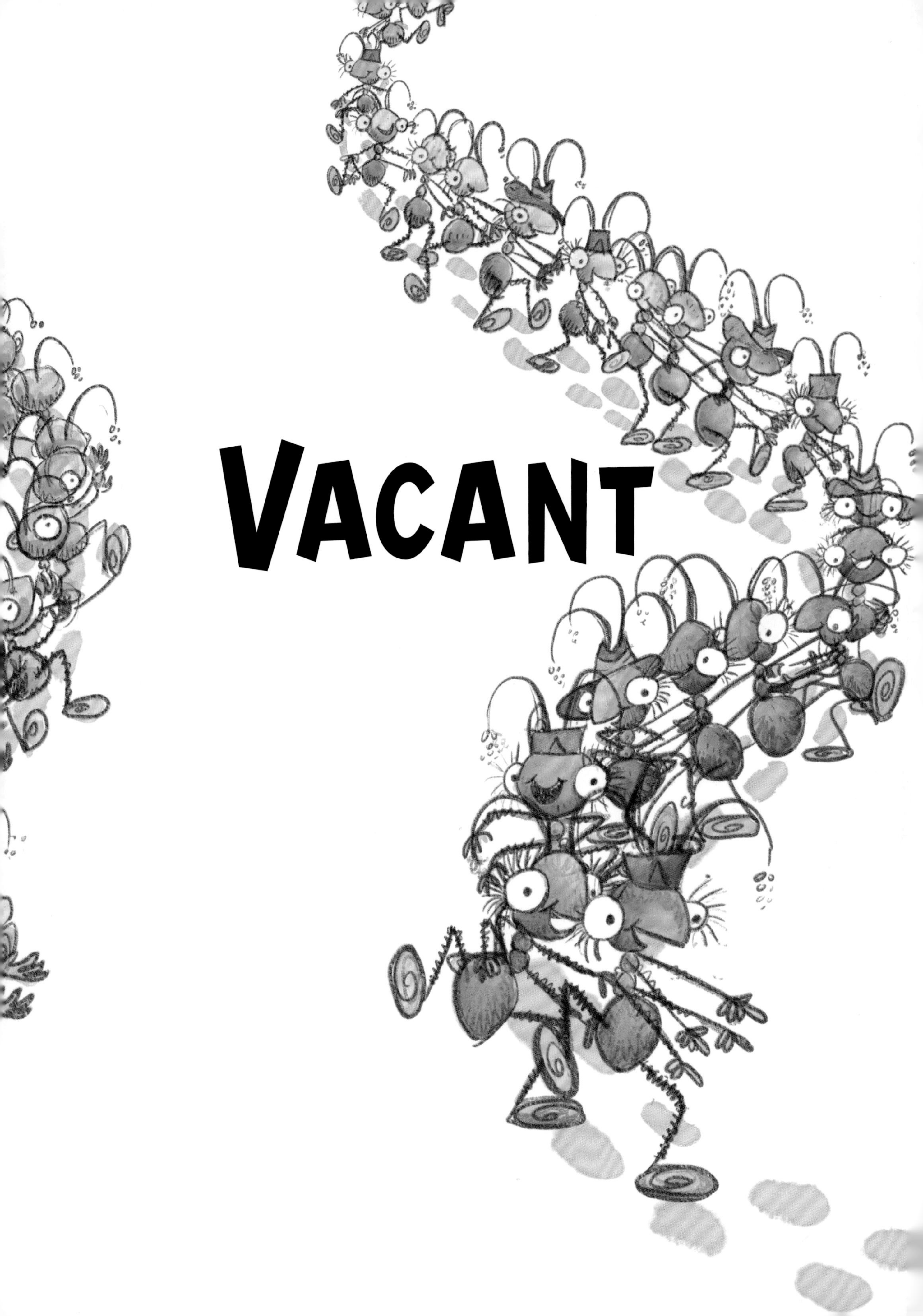
VACANT

ANT

BRILLI-

-ANT!

AMAZING FACTS ABOUT ANTS

Did you know that there are more than 12,000 different ant species? They can be found on every continent in the world, except ANTartica!

Researchers estimate the total number of all the ants in the world at 20 quadrillion —that's 20,000,000,000,000,000—which works out to 2.5 million ants for every human being on the planet!

The eggs laid by the Queen hatch into worm-shaped larvae with no eyes or legs. Nursery ants provide them with a constant supply of food, to fuel their rapid growth. When the larvae reach the right size, they undergo metamorphosis into a pupa. At this stage they look more like an adult, but their legs and antennae are not ready to use.

Eventually the pupa emerges as an adult, at which stage it is fully grown. The process from hatching to adulthood can take several weeks.

Ants have a specific role within their colony, and that role may change over time. Every colony has one or more egg-laying Queens. Younger female ants (called nursery or worker ants) stay in the nest and look after the Queen and her babies. Soldier ants defend the nest, while older worker ants forage for food. Male ants have wings and their only role is to mate with the Queen to ensure there are plenty of babies so the colony will continue.

Did you know that ants don't have ears? There are a few species who don't have eyes either! They listen via vibrations in the ground, and can send chemical signals—called pheromones—through their body to signal other ants. They'll do that to warn that danger is near, or to create a trail that will lead the other ants to a food source. Think of Hansel and Gretel using breadcrumbs to mark the way home—ants do a similar thing, but with smells!

(AND ELEPHANTS!)

Did you know that ants are fearsome warriors? There's even a species, the Dracula ant, that holds the record for the fastest movement in the animal kingdom. This incredible ant stuns its prey by snapping its jaws shut at a speed of up to 320km per hour, 5000 times faster than the blink of an eye! The fastest land animal—the Cheetah—has a top speed of just 120kms per hour!

Ants are also incredibly strong and can lift up to 50 times their own body weight. That's like you lifting ten baby elephants all at once! The common American field ant is reported to have a neck joint that can withstand pressure up to 5000 times the ant's own weight.

Did you know that ant colonies are also described as superorganisms, because the ants work collectively to care for and support their colony? They share labour, communicate with each other and are able to solve complex problems together. Ants do such a good job that some Queen ants live for up to 30 years!

Obviously, ants really are BrilliANT! If they were humans, we'd probably consider them to be superheroes!

There's even a special name for the study of ants. Myrmecology (mur-mi-KOL-uh-jee) is the branch of insect science (Entomology) that studies Ants. Perhaps you might become a Myrmecologist one day!

By the way—if you grow bananas at home, look out for elephants! They really love bananas and won't hesitate to trample everything in their path to get to their favourite snack—even you!

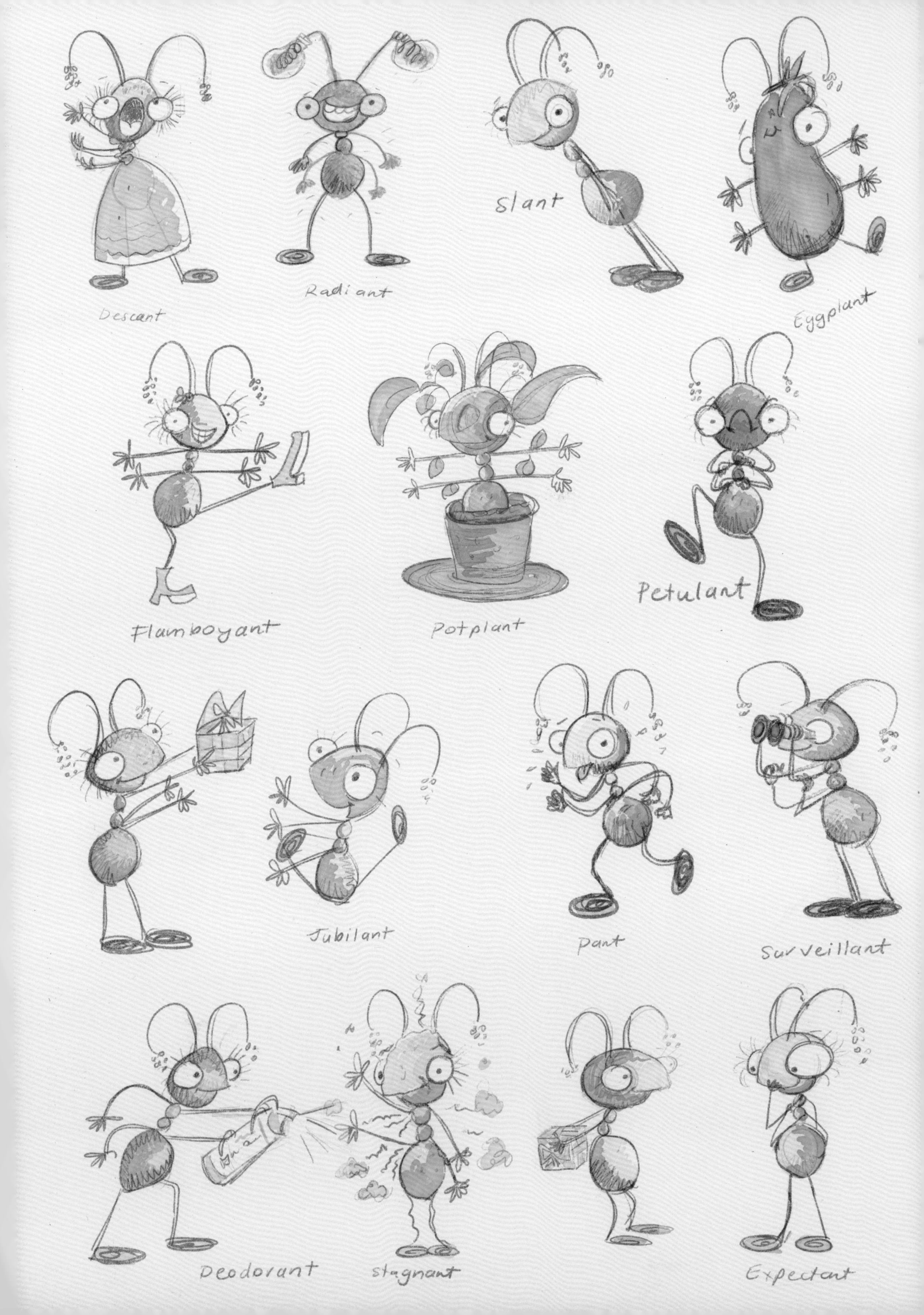
Descant
Radiant
Slant
Eggplant
Flamboyant
Potplant
Petulant
Jubilant
Pant
Surveillant
Deodorant
Stagnant
Expectant